Our World

Oil

By Kate Bedford

Aladdin/Watts
London • Sydney

© Aladdin Books Ltd 2006

Designed and produced by
Aladdin Books Ltd
2/3 Fitzroy Mews
London W1T 6DF

553.282

First published in 2006 by
Franklin Watts
338 Euston Road
London NW1 3BH

Franklin Watts Australia
Hachette Children's Books
Level 17/207 Kent Street
Sydney NSW 2000

A catalogue record for this
book is available from the
British Library.

ISBN 0 7496 6278 6

Printed in Malaysia

Editor:
Harriet Brown

Designer:
Flick, Book Design and Graphics

Consultants:
Jackie Holderness – former Senior Lecturer in
Primary Education, Westminster Institute,
Oxford Brookes University

Rob Bowden – education consultant, author
and photographer specialising in social and
environmental issues.

Illustrations:
Ian Thompson

Picture researcher:
Alexa Brown

CONTENTS

Notes to parents and teachers	4
What is oil?	6
How oil is formed	8
Black gold	10
Finding oil	12
Drilling deep	14
From oil-field to refinery	16
Uses for oil	18
Oil and the environment	20
Cleaner oil	22
Running out	24
Saving oil	26
Oil for the future	28
See how much you know!	30
Key words	31
Glossary	31
Index	32

Notes to parents and teachers

This series has been developed for group use in the classroom as well as for children reading on their own. In particular, its differentiated text allows children of mixed abilities to enjoy reading about the same topic. The larger size text (A, below) offers apprentice readers a simplified text. This simplified text is used in the introduction to each chapter and in the picture captions. This font is part of the © Sassoon family of fonts recommended by the National Literacy Early Years Strategy document for maximum legibility. The smaller size text (B, below) offers a more challenging read for older or more able readers.

Running out

Oil took millions of years to form. Once it is used, it cannot be replaced. One day it will run out.

A

◄ This new car uses much less petrol.

New hybrid vehicles have been invented which use two or more sources of power.

B

4

Questions, key words and glossary

Each spread ends with a question which parents and teachers can use to discuss and develop further ideas and concepts. Further questions are provided in a quiz on page 30. A reduced version of pages 30 and 31 is shown below. The illustrated 'Key words' section is provided as a revision tool, particularly for apprentice readers, in order to help with spelling, writing and guided reading as part of the literacy hour. The glossary is for more able or older readers. In addition to the glossary's role as a reference aid, it is also designed to reinforce new vocabulary and provide a tool for further discussion and revision. When glossary terms first appear in the text they are highlighted in bold.

 ### See how much you know!

What is crude oil?

How is oil made?

Can you name some of the things oil is made into?

How do geologists find oil?

Why does oil pollute the environment?

How is oil taken out of the ground?

What are hybrid vehicles?

How many ways of saving oil can you think of?

Key words

Crude oil

A

Drill bit

Fossil fuel **Geologist**

Oil rig **Petrol**

Plankton **Rock**

Refinery

Glossary

Crude oil – Oil that has come straight from the ground.
Fractionating tower – The tower at a refinery in which oil is separated into different substances.
Microbes – Tiny plants and animals that are so small that they can only be seen through a microscope.
Oil-field – An area that naturally contains a lot of crude oil.
Pollution – Harmful chemicals, gases or rubbish that damage the environment.

B

Reservoir – A chamber that stores liquid. Crude oil is contained in underground reservoirs.
Sediment – Tiny pieces of rock and soil that are carried in sea-water.
Slick – A layer of oil that forms on water after an oil spill.

What is oil?

Oil is one of the world's most important resources. We use oil for energy. We use oil to power cars, trucks, planes, trains and ships. Oil is also used as a raw material to make many things. Every day, we use things that are made from oil.

► Crude oil is thick and sticky. It is made into fuel.

Oil is found underground. When it comes straight from the ground, we call it **crude oil**. There are several different types of crude oil. They vary in colour, smell and thickness. Heavy crude oil is dark brown or black. It is very sticky like tar. Light crude oil can be clear in colour and is very runny.

Crude oil is used to make all these things.

Plastics

Oil is made into many different kinds of plastic. Foamy plastics, with bubbles of air blown into them, are used to fill furniture.

Disposable nappies

One cup of crude oil is used to make the plastic for a disposable nappy.

Cosmetics

Crude oil makes people look beautiful. Many cosmetics are made from oil.

Paint

Paint can also be made from oil.

Cloth

Many strong, long lasting, **synthetic** materials, such as nylon, are made from oil.

Bitumen

The black sticky substance called bitumen, which sticks the stones together on our roads, is made from oil.

Detergents

The detergents we use to wash clothes are made from oil.

 What things have you used today that are made from oil?

How oil is formed

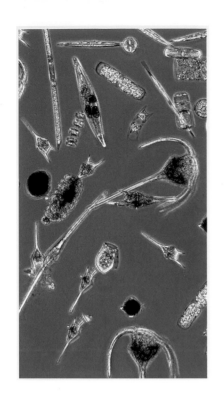

Oil is a fossil fuel. It is made from the remains of tiny plants and animals called plankton. The plankton lived in the seas when dinosaurs still ruled the world. Over millions of years the tiny plants and animals have changed into oil.

▲ ▶ **These tiny plankton float in the sea.**

Tiny plankton living in our seas today are similar to the prehistoric plankton from which oil is made. Like today's plant plankton, the tiny prehistoric plants used energy from sunlight to make their food. When they were eaten by animal plankton, the Sun's energy passed into the tiny animals. So, oil made from prehistoric plankton is actually stored energy from the Sun.

It takes millions of years for oil to form.

1. Millions of years ago plankton died and fell to the seabed.

2. Layers of dead plants and animals built up on the seabed.

3. Mud and sand (**sediment**) sank to the bottom of the sea and covered these layers of plankton.

5. Slowly the muddy sediment turned into rock.

6. The plankton rotted, giving off bubbles of gas, and turned into a thick liquid. This is crude oil.

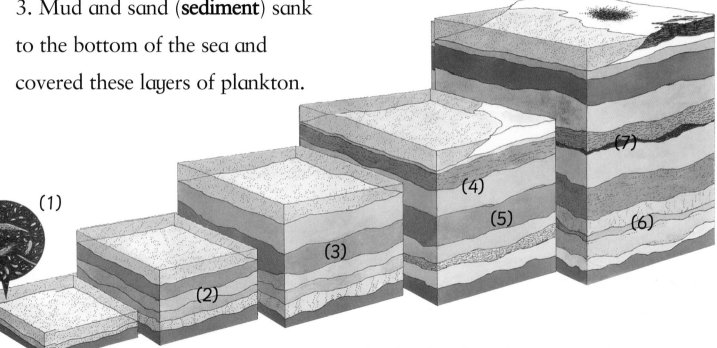

4. More layers of sediment piled up on top of the dead plankton. The bottom layers were squashed and became hotter.

7. Crude oil and gas seeped up through the spongy rock, until they reached a solid rock. They stayed under the solid rock in a pool or **reservoir**.

 Which ingredients do you need to create oil?

Black gold

Every day millions of people use oil in some way. Yet, it is only in the past 150 years that oil has become so important to so many people. Oil is so valuable that it is sometimes called 'black gold'. Imagine life without plastic and fuel.

◄ **Oil was once used in lamps.**

Oil has been used for thousands of years. In places where oil naturally seeped to the surface, people collected it. The Ancient Babylonians used it as a mortar for building. The Native Americans used it as a glue. In the 1850s, people decided they would not wait for the oil to come to the surface. They would drill for it.

▼ When cars were invented, more oil was needed.

When the petrol engine was invented the demand for oil grew. Cars became more common. Now, oil has become a vital part of our life. Many **oil-fields** have been discovered and about 80 million barrels of oil are produced each day.

Wars are fought over oil.

Oil has become so important and valuable that wars have been fought over oil-fields. In 1991, Iraq invaded its neighbour Kuwait because it wanted Kuwait's oil-fields. The UK and USA went to war with Iraq to help free Kuwait. Many of the oil-fields were deliberately set on fire by the Iraqi soldiers. Burning and leaking oil caused a lot of **pollution** to the environment.

 How did people travel before cars were invented?

Finding oil

Oil is very valuable. Geologists, people who study rocks, look for clues that show where oil might be in the ground. They study the shape of hills, the colour of rocks, and the type of soil, to find out if they might contain oil.

▶ **This geologist is looking for oil.**

Geologists search for rocks that might have oil in them. They can do this by sending sound waves and electric currents through the ground. Geologists can work out where there might be rocks with oil in them by studying how fast the waves and currents travel through the rock.

◀ Rock samples are drilled out of the ground.

If geologists find rocks that could contain oil, they drill a small test well. A sample of rock called a 'core' is brought up to the surface. The rock sample is checked for oil. If there is oil in the rock, the geologists will see tiny droplets of oil clinging on to the rock, a bit like drops of water clinging on to a window after a rainstorm.

Oil is found all over the world.

Oil is found in many countries all over the world. It is found on land and under the sea. The Middle East has two thirds of the world's known oil reserves. The amount of oil in an oil-field is measured in barrels. Saudi Arabia, in the Middle East, has 262 thousand million barrels of oil.

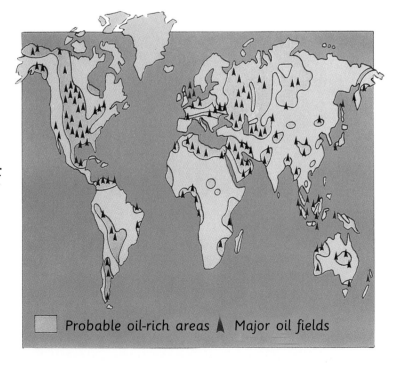

Probable oil-rich areas ▲ Major oil fields

 Can you name any different types of rock?

Drilling deep

Drilling rigs are used to drill into the ground to reach oil. The rigs have a flat base called a platform. A tower holds up the drilling equipment. The drill is a steel pipe which has a point at the end. This is called a 'drill bit'.

▶ Oil-rigs can be built in the sea.

When oil is found under the seabed, an oil-rig is built over the oil-field. Different rigs are used depending on how deep the water is. Oil-rigs with legs are used in water less than 400 m deep. Floating rigs are used in water up to 1,000 m deep. They are tied to blocks on the seabed to stop them drifting away. In even deeper water, rig ships are used, which are kept in place by anchors.

◀ The drill bit cuts down through the rock.

As the drill bit bores deeper into the ground, more lengths of pipe are added at the top. The drill bit (below right) is covered with diamond dust. Diamonds are very hard and strong, so they help the drill to grind through the rock.

Drills reach oil deep under the ground.

In many oil-fields, the oil is pressed and squashed deep inside the rock. When a hole is drilled into the rock, the pressure pushes the oil upwards and out. The pressure can last for days or years, but eventually the flow of oil drops. Then, pumps are needed to bring more oil to the surface.

 Why is diamond dust used to cover the cutting drill bits?

From oil-field to refinery

Oil is often found in remote places. It is drilled in the middle of seas, hot deserts and frozen lands. After the oil has been brought to the surface, it is taken from the rig to a refinery. In the refinery it can be turned into many useful products.

 **Oil is carried along pipes or in huge ships.**

Oil is transported to refineries by pipelines. Pipelines run along the seabed or across land. Overland pipeline networks run for thousands of kilometres across North America.

Oil travels by sea in huge tanker ships. These ships are nicknamed supertankers. They are as long as four football pitches placed end to end!

Oil refineries are like huge factories.

At a refinery, crude oil is separated out into different substances. These can be used for many different things.

The oil is separated out in this tower (below).

At the refinery, heated crude oil is fed into a tall tower called a **fractionating tower**. The tower is coolest at the top and hottest at the bottom. The crude oil separates into different substances. The lightest gases rise to the top of the tower and the heavier substances are below. Products such as kerosene and petrol are piped off at different levels within the tower.

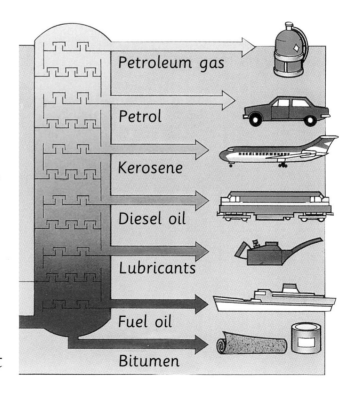

Petroleum gas

Petrol

Kerosene

Diesel oil

Lubricants

Fuel oil

Bitumen

 Why are oil refineries often built on the coast?

Uses for oil

Around 85 per cent of refined oil is made into different types of fuel. This fuel is used to power vehicles. It is also used to heat our homes, schools and workplaces. In power stations oil is used to produce electricity.

► **Oil is used to power these vehicles.**

We use one third of the world's oil to make fuel for transport. Oil burns easily and is a good source of energy. Crude oil is used to make fuels such as petrol, diesel and aviation fuel. These fuels provide the energy that powers cars, trucks, ships, trains and planes.

▶ Oil keeps car parts moving smoothly.

Vehicles and machines need oil to keep their moving parts working. Grease, brake fluid for the brakes and gear lubricants are all made from crude oil. Using lubricant oils prevents wear and tear on moving parts and makes machines last longer.

Oil can be used to make electricity.

We burn oil in power stations to make electricity. When the oil is burned it releases its energy as heat. The heat boils water and makes steam. The steam then turns turbines which generate electricity.

 In how many ways have you used oil energy today?

Oil and the environment

Oil gives us energy and many other things but it also causes pollution. Oil pollution makes the environment dirty. It can harm and kill wildlife and people.

 Spilt oil can harm wildlife.

If oil is spilt, it can cause serious damage to the environment and kill many birds and animals. Oil spills from oil tankers at sea are especially harmful as the oil forms a thick, black layer called a '**slick**'.

The slick clogs birds' feathers and the fur of marine mammals, such as otters. Animals can also swallow oil as they try to clean themselves. In the past, large oil spills have destroyed entire coastlines.

► Burning fuel made from oil causes smog.

Vehicles usually burn diesel or petrol. Their exhaust pipes release fumes and gases such as sulphur dioxide. Diesel also gives out tiny specks or particles like soot. In calm sunny weather, these substances collect over cities and form smog.

Burning oil is making the Earth grow warmer.

When we burn oil, large amounts of carbon dioxide gas are released into the air. Normally, carbon dioxide is absorbed by plants, but human actions are releasing too much carbon dioxide. This forms a blanket that traps the Sun's heat in the air around the Earth. This is causing temperatures on Earth to rise – a process known as global warming.

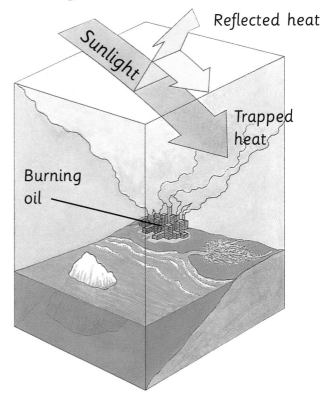

Reflected heat

Sunlight

Trapped heat

Burning oil

 Which cities have the worst smog pollution?

Cleaner oil

Oil is being made cleaner to help protect the environment. Harmful chemicals can be taken out of oil so it pollutes the environment less. Scientists are also finding safer ways to carry oil. They want to prevent oil spills and protect wildlife.

▶ **New oil tankers are safer than old tankers.**

Oil tankers carry millions of gallons of oil. If they become damaged in an accident, they can cause huge oil spills. Nowadays, tankers can be built with double hulls. This means that if only the outside hull is damaged, the oil is still held safely inside the inner hull. Soon only double-hulled oil tankers will be allowed to carry oil.

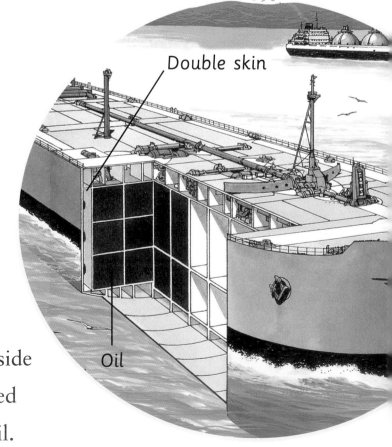

Double skin

Oil

Cleaner oil is being used to heat our homes.

We can use cleaner heating oil in our boilers to heat our homes. It has up to 90 per cent less sulphur in it than normal heating oil. This new oil burns cleaner and more efficiently. Cleaner oil helps to reduce pollution.

This sludge is left at the bottom of oil containers.

Oil sludge is the thick, sticky substance left at the bottom of oil tankers and storage containers. It is difficult to get rid of. In the past it has been disposed of in landfill sites or burned, but this causes pollution. A new method has been invented which cleans the sludge with steam. This separates the oil from the sludge. This recycled oil is then used to power ships.

 What materials can you recycle?

Running out

Oil took millions of years to form. Once it is used, it cannot be replaced. We are using it up quickly and one day it will run out. Think about all the things we use oil for. What will happen once the oil has run out?

▶ **This graph shows how oil may run out in the future.**

Oil is non-renewable. This means that it cannot be replaced. Once it has been used up, it is gone forever. About half the world's oil resources have already been used. Oil is expected to run out in about 40 years' time. In the future, if no more oil is found, less and less oil will be produced. The oil that is left will be much more expensive.

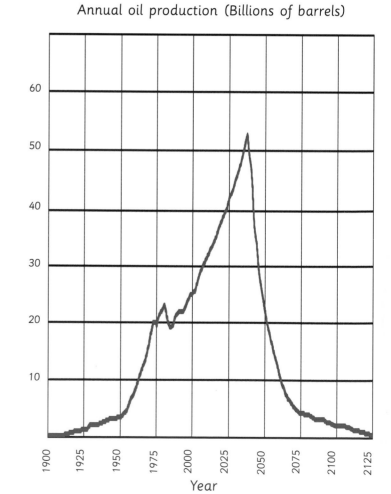

Annual oil production (Billions of barrels)

Year

Without oil, today's cars could not run.

Imagine what life would be like without oil. Most forms of transport would stop. Fewer cars, buses, trains, planes or ships could run. This would have a huge effect on our modern life because we rely on oil to power most of our transport.

▲ **Many people in cold areas rely on oil to heat their homes.**

Many people live in places that have very cold winters. Heating their homes by burning oil keeps them warm. When the oil runs out, many people may have problems heating their homes. They will need to use another source of fuel.

 How would your life change if there were no more oil?

Saving oil

We can all help to save oil. If we use less oil, the oil we do have will last longer. Oil is an important source of energy. There are many things we can do to use less energy.

▶ **Hybrid cars use much less petrol.**

New hybrid vehicles have been invented which use two or more sources of power. Some of these cars use petrol, and electricity from batteries. The batteries recharge themselves as the car drives along. The car switches from using electricity to petrol and back depending on its speed. Hybrid cars produce much less pollution and are very fuel efficient.

Why not try some of these ways to save oil?

Use less energy at home. Put on a jumper instead of turning on the heating. Houses with insulated attics and lofts save 25 per cent more energy.

Save petrol by using a bus or train instead of going in the car. Walking and cycling keep you fit and save oil.

Save electricity. Remember oil is used to generate electricity. Switch off lights when you leave the room. Don't leave the television on standby. It uses up lots of electricity on standby. Use low energy light bulbs. They use a quarter of the electricity used by ordinary bulbs.

Recycle steel and aluminium. This uses much less energy and saves raw materials. So recycle your empty drinks cans or baked bean tins.

Recycle your plastic bottles and bags. They are made from oil and can be reused. They can even be made into fleeces to keep you warm.

 What are you going to do to save oil?

Oil for the future

Scientists are working on new ways to power the world in the future. New inventions will help us to make the best use of oil. Cars that run on both petrol and electricity will become more common and help to save oil.

▶ **This rock can be used to make oil.**

Oil shales are rocks that contain a large amount of organic matter (remains of dead animals and plants) called kerogen. Kerogen can be made into man-made oil. Oil shale is found in many parts of the world. To change kerogen into oil it must be heated to high temperatures. The problem with making oil from oil shale, is that it uses lots of energy to heat the kerogen, and it produces large amounts of waste material.

▶ More oil reserves may be hidden under the ground.

Oil companies are constantly searching for new oil reserves. A large oil reserve was recently discovered in Bohai Bay Basin off the north coast of China. In the future it is believed that more oil may be discovered in Russia, the Middle East, north and west Africa and central South America.

These tiny microbes may be used to get oil out of rocks.

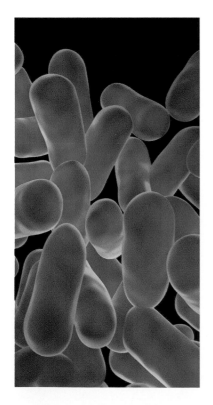

In many oil reservoirs, three quarters of the oil is left behind because it is too difficult to get out. Scientists are working on ways to get the rest of the oil out. One way is to use **microbes.** These are tiny animals which feed on the nutrients in the oil. They release gas. The gas builds up the pressure in the reservoir which forces more oil out of the rock.

 How will using electric cars help the planet in the future?

29

See how much you know!

What is crude oil?

How is oil made?

Can you name some of the things oil is made into?

How do geologists find oil?

Why does oil pollute the environment?

How is oil taken out of the ground?

What are hybrid vehicles?

How many ways of saving oil can you think of?

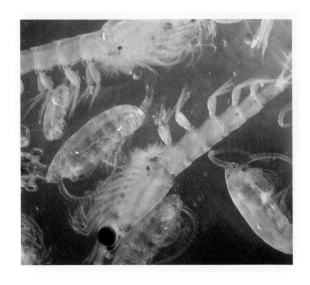

Key words

Crude oil

Drill bit　　**Energy**

Fossil fuel　**Geologist**

Oil rig　　　**Petrol**

Plankton　　**Rock**

Refinery

Glossary

Crude oil – Oil that has come straight from the ground.

Fractionating tower – The tower at a refinery in which oil is separated into different substances.

Microbes – Tiny plants and animals that are so small that they can only be seen through a microscope.

Oil-field – An area that naturally contains a lot of crude oil.

Pollution – Harmful chemicals, gases or rubbish that damage the environment.

Reservoir – A chamber that stores liquid. Crude oil is contained in underground reservoirs.

Sediment – Tiny pieces of rock and soil that are carried in sea-water.

Slick – A layer of oil that forms on water after an oil spill.

Synthetic – Man-made. Plastic is a synthetic material.

Index

C
carbon dioxide 21
cars 6, 11, 18, 19, 25, 26, 28, 29
cleaner oil 22, 23
crude oil 6, 7, 9, 17, 18, 30, 31

D
diesel 18, 21
drill bit 14, 15, 31
drilling 10, 13, 14, 15

E
electricity 18, 19, 26, 27, 28
energy 6, 8, 26, 27, 28, 31
environment 11, 20, 22, 30, 31

F
finding oil 12-13
fossil fuel 8, 31
fractionating tower 17, 31
fuel 10, 18, 21, 25

G
geologists 12, 13, 30, 31

H
heating 18, 23, 25, 27

L
lubricants 19

M
microbes 29, 31

O
oil-fields 11, 13, 14, 15, 16, 17, 31
oil refinery 16, 17, 31
oil reserves 13, 29
oil-rigs 14, 16, 31
oil shale 28
oil spills 20, 22, 31

P
petrol 11, 17, 18, 21, 26, 27, 28, 31
pipelines 16
plankton 8, 9, 31
plastics 7, 10

pollution 11, 20, 21, 22, 23, 26, 30, 31
power stations 18, 19

R
recycled oil 23
refined oil 18
reservoir 9, 31
rock 9, 12, 13, 15, 28, 29, 31
running out 24

S
saving oil 26, 27, 28, 30
sediment 9, 31
ships 6, 14, 16, 23, 25
slick 20, 31
synthetic 7, 31

T
tankers 16, 20, 22
trains 6, 18, 27
transport 18, 25

W
wildlife 20, 22